A Tangled Quandary

Brenda Ramsey Eagle

Copyright © 2020 Brenda Ramsey Eagle
All rights reserved
First Edition

PAGE PUBLISHING, INC.
Conneaut Lake, PA

First originally published by Page Publishing 2020

ISBN 978-1-64628-951-6 (pbk)
ISBN 978-1-64628-952-3 (digital)

Printed in the United States of America

A TANGLED QUANDARY

CHAPTER 1

THE BEGINNING

I was born in Denver, Colorado, the only one of my family not born in Oklahoma or Arkansas. My mother and dad, pregnant with me, decided to move to Denver, where some family were, so they could get a fresh start together. Then I was born, and we lived there awhile before moving back to Arkansas where my parents settled and had three more children, two sisters and my youngest brother. I don't remember anything about Denver, as I was so young, but I am going there someday soon.

My dad worked as a gas station attendant until he and a couple of friends took welding classes and began making metal trusses, support beams for chicken house construction, their own business. This was a very from-ground-up kind of business as they started it with welding equipment on the back of a big truck, under a shade tree. Then going to a building and making office in my parents' and Dad's partners' homes, they ran it for several years and, after getting good amount of clients, sold it to a local company, and my parents bought a farm. I was a very blessed baby girl born into a family of what I like to call Pioneer women: six grandmothers, great-grandmother, and great-great-grandmother, and including my mother. All my grandmothers were married and had several children. My grandfathers, most of whom I did not know because they all died at middle age, were very hard working men, not having all the modern-day conveniences—like bringing water up with a bucket from well tied with a rope, pitchfork loose hay to cows instead of baled, horse-drawn plow led by Grandfather to turn the dirt for planting. But after losing their husbands, my grandmothers never remarried. When I was old enough to notice, I asked a couple of them why they never remarried, and they said, "I loved your grandfather, have had my children, and will continue to raise them on my own." That they did, and very well. We had huge gardens full of delicious vegetables, garden tilled by horse-drawn plow or single-engine tractor. We raised all our own cows, chicken, and pigs. Dad, grandfathers, uncles, and cousins would process, cure, and prepare meat for eating and storing. The women processed the chickens after killing by heating water for dipping the bird into it so feathers could be plucked by hand, then cleaned and cut up for freezer storage. We then, men and women, would harvest the "huge" garden—beans green and brown, corn, tomatoes, cabbage, lettuce, and much more, depending on available seed or plants.

Next the processing would start: snapping and shelling beans, shucking corn, cleaning and chopping the tomatoes and cabbage for canning. Things like tomato juice, salsa, "cowboy caviar" made with black-eyed peas, onions, green peppers, with olive oil, vinegar, sugar, garlic, and salt mixed and poured over. Sauerkraut made from cabbage, corn, and green beans. The best part was the downtime when we would celebrate with a big meal of fresh meat and vegetables—fried chicken roast with new potatoes, pot of brown beans, corn on the cob, corn bread, or homemade rolls, etc. Not to mention, milk from our jersey cow, used to make butter, to use with sorghum and molasses, a sweet substance like syrup but much more nutritious. Of course, especially desserts with homemade pie shells for fruit pies and cobblers. All the kids would bring jars, with vented lids, in the spring and summer to catch lightning bugs. We could keep them in our room for the night to light up after lights off; we would let them go and catch new ones the next night.

June bugs would also come out in June and July, during garden and fruit harvest. We could catch them tie a sewing string to one back leg, which requires a lot of patience from their wiggling. We then would let them fly and hold the string.

A TANGLED QUANDARY

Another nice thing was all our family getting together, including both my mother and my dad's parents, etc., and the work and rewards were shared. Also we did not have to go to several dinners to see everyone on the holidays as everyone met at one place.

Here is little BRENDA LEE RAMSEY, 14 months old daughter of Mr. and Mrs. Berlin Ramsey of Siloam Springs, a lucky little girl with her six grandmothers. Left to right, her maternal grandmother, Mrs. Frieda Arnold of Lincoln; her maternal great grandmother, Mrs. Lena Hatchette of California; her maternal g r e a t-great grandmother, Mrs. Merinda Arnold of Lincoln; her paternal great grandmother, Mrs. Leona Ramsey of Swain; her maternal great grandmother, Mrs. Ethel Arnold of Summers; h e r paternal grandmother, Mrs. Delia Ramsey of Watts, Oklahoma. The great great grandmother is 82 years old.

BRENDA RAMSEY EAGLE

CHAPTER 2

Ramsey's Bait Shop

During the summer break, my sisters and I would go to Dad's parents in Oklahoma for a week. In Watts Oklahoma across from the lake and dam for wonderful fishing. We would help them in cleaning the store, waiting on customers, and sold bait for fishing. Next we would help them with raising worm and netting for minnows out of nearby creek to fill up their inventory. If you have never netted for minnows, which are like anchovies, we would wear leg-long rubber-boot-like pants; take a big net, like the fishermen on television; and put crackers in the water. When minnows came to get food, we would catch them in net and put in water tank at bait store, which had an electric motor for oxygen. Also another way of catching the minnows was to use a two-gallon glass jug which had flumed holes in the sides so minnows could go in, but the holes on inside was much smaller, and they could not get out. We also put the saltine crackers in the jug to lure them in. The worms weren't as much fun; they were raised from original worm stock in a dirt mix of rabbit droppings, mixture of meal and fine grain. In a confined area with boarded rails and ground dirt then we picked up and counted, put into Styrofoam cup with vented lid on, and stored in refrigerator until purchased. It was a thriving business. As well as bait and tackle for fishing, they had rolls of fresh lunch meat and cheese, which made the best sandwich. Along with ice-cold soda pop and chips, there were many to choose from on a hot summer day. I remember an instance when

Grandma and my sister went to the house to fix lunch; their house on the hill was in the back of store.

While they were gone, I was taking a nap on bread rake—they had made a little daybed for us out of customers way—when several loud people came in. I looked at them, and they were dressed different: ladies in dresses, with bandanas on head, sandals; men with vests and bright-colored shirts. They were much different from people we were used to seeing. I got behind the counter with Grandpa, and they picked up a few items a lady brought to the counter, and she then pulled the top of her dress down, exposing herself. Grandpa froze.

I said, "Grandpa, they are taking things out and not paying for them."

He told me to get down, reached for a little gun, and said, "Stop, I will call the police."

The ones still in the store dropped what they had and were all on bikes and truck, quick to leave. When we told Grandma, she said, "Good thing I wasn't there, exposing herself would have been least of her worries." Grandma was a tough woman. The experience definitely made an impression on me that people determined to do bad things would find away, but my Grandpa made his point without casualty.

Grandpa loved making people laugh and providing for my grandmother. After dinner he would take who wanted to hear one of his stories out on front porch, and we would listen for as long as he would talk. And it was no surprise that after he passed away, my grandmother received a $300 check from a November 1982 edition of *Reader's Digest* where he had submitted this story:

A farmer known for his thrift moved to an area often ravaged by tornadoes. He was unaccustomed to such violent weather, and stories about past disasters made him so apprehensive that he parted with some of his money to build a storm cellar. Every time the sky darkened and the winds whipped up, he'd run into the cellar, and every time he'd come back up to find no evidence of a storm. Soon he started to feel that he'd wasted his money. Then one day he emerged from his cellar to find that a tornado had leveled his home and barn. Debris was everywhere. The farmer took one look around, glanced skyward, and bellowed, "Now that's more like it!"

A TANGLED QUANDARY

CHAPTER 3

FARMWORK

Growing up as a daughter of farming parents, my dream was always to find a God-fearing man who wanted to enjoy the farming life, get married, and have a house full of children—but I certainly did not get my dream. I love my family, school, friends, and country life. The sounds of frogs croaking in nearby ponds, crickets, jarflies also called cicadas, are such refreshing noises compared to the loud traffic in big cities. Something about the cicadas bugs that are so much fun and interesting is that they come out from the ground after two to seventeen years as nymphs, then shed their skins that look like a strange light-brown replica of the adult, which they leave behind and fly away. The jarfly skeleton can be taken carefully from the tree trunks, fence posts, etc. and put in container for show-and-tell at school.

We always had cattle and pigs, gardened, and later became chicken farmers. We even had a dairy for a while, which was definitely different from raising chickens. It started by getting up at 4:00 a.m., and that time of the morning gave me my first taste for hot black coffee. We would round up the Holstein cows, which were usually already lined up at the milk barn door for their full grain breakfast and milk swollen relief. Hooking up the milking devices did not take long, but it required patience waiting for the empting. Sometimes the cows can be irritable and hard to get hooked up. One time I had trouble getting mine to settle down, and she kicked the milker out of my hand and cut my forearm with her sharp hoof. Dad sent me to the house for wound-cleaning and so Mom could get the bleeding stopped.

After cleaning each milking device and clearing all the cows from the barn, it was time to get cleaned up for school bus. The summer was full of cutting, racking, baling hay, selling chickens or getting ready for more. Cleaning out the chicken houses to get ready for baby chicks was Dad's job, but after the manure was scraped off the chicken house floor with tractor blade, it was Mom's and kids' turn to scatter straw or rice hulls. Next the cardboard boxes would come on a pallet, flat, and had to be formed into four-by-four squares for chick feed. The water jars were glass with plastic lids and had to be washed and sanitized then cardboard feed boxes, and jars were placed around stoves, six per stove by ten to twelve per house. We had two houses until a tornado took half of one. My dad was in the chicken house when it hit. Luckily, he made it to the half that was still standing. He then made it into a hay barn. It was then my sister and I's job to wash all the glass water jars and lids for the next batch of baby chicks. I remember my sister and I would race to see who would be done first, and I always beat her and would then help her finish. So one day I decided to sneak to her chicken house and see why I would always beat her. She was filing her nails, and I realized it was her way of getting me to help her.

We had a lot of close calls with tornados living on Oklahoma-Arkansas border, but praise God, no one ever got hurt. I remember several times we were in the house, and my parents watching weather news would say, "Kids, time to go to the storm shelter/root cellar underground," where we stored our fresh canned food. When we were heading out of the house, my mother would hand us pots and pans to put on our heads as the hail was big and scary—I have to say, it was also very loud under metal pot. After getting the houses ready for chicks and hay racked ready for baling, we had two nice ponds full of fish, for fishing, riding horses, and riding bike were our playtime. We also had a very special family time after my parents bought a reel-to-reel tape recorder. My Mother would play music and sing while we were at school and after dinner we would all gather around the dining table to sing. Dad was very good on the harmonica and guitar and also had a good singing voice for the old country music and especially when he and Mom would sing together. My three sisters and I would sing also but were better at telling jokes, and my little brother was so cute repeating our jokes. Also when we had family gatherings, we would hide the recorder under the couch and put the microphone out to record conversations and fun things that would happen. We had a whoopee cushion—in case you haven't heard of it, it blows up like a balloon and has a flat nozzle for letting out air slow, and when put under couch cushion and an unsuspecting person sits on it, those around think…and I will leave the sound to your imaginations. This, along with the laughter, is so cool to hear now with my dad and several grandparents already passed. This was so very important to me, and when I left home, I took the reels with me, and when I traveled, I tried to get them recorded onto cassettes or CDs. In the meantime, my mother's house burned, and the recorder was in the house. I was so very thankful I had thought to take the reels with me. I finally was able to get them recorded at a new studio in Fayetteville, Arkansas, and gave them to my sisters, brother, and their kids. It was very important at my dad's funeral as he sang his own songs from the cassette and played the harmonica; we will do the same for Mother. I have also been the -photographer in my family. There were thousands at least; my walls are not covered with expensive paintings but many framed pictures of my loved ones and special events. These are so comforting to me, and now I am so proud to say that my niece, of my brother, has taken a serious interest in -photography, not only for herself but for others' weddings, graduations, births.

My dad has one brother who has three girls, and they would come visit us in the summer. We would adventure across fields and up steep hills with streams to play in and huge moss-covered rocks to climb on…where my high school sweetheart and I would later carve our names in a tree.

Once, in particular, when our cousins were visiting, we found a burnt pile of leaves, and hearing the train horn in a nearby town, Westville, Oklahoma, we were sure it was a hobo traveling through. We ran all the way home. We also had lots of beautiful deer that would shed their horns and arrowheads to search for.

A TANGLED QUANDARY

We also lived near Harrison, Arkansas, where there was a big amusement park called Dogpatch. It had a small roller coaster and water boat rides, delicious food, kettle corn, cotton candy, ice cream treats, hamburgers, hot dogs, and much more, along with booths full of handmade and glass-blown items. This was my parents' vacation for us, and my grandparents would pack a picnic of fried chicken, sweet ice tea, dessert, and we would eat before entering the park. My last memory of there is my grandfather on the roller coaster with us and seeing the smile of joy on his face.

In my middle school years, Dogpatch closed, and we heard of another amusement park, which is still going and much bigger called Silver Dollar City, in Branson, Missouri. Last year my sister heard that there was a crafts fair on the grounds where Dogpatch was, so my mother, niece, sisters, and I went to see what might be left. There was only a few of the buildings where glass-blowing and basket-making were, and the slide for water boat ride, sad to say, it all gone and run-down, but we were able to share memories with my niece.

Another vacation I remember was my whole family in our brown panel station wagon going to lake for the weekend. We would take sleeping bags, and our car had a lot of room for sleeping. This was a trip before my brother was born, just my sisters and me. Dad would park us near the water where we could build a nice campfire. Mom made us our favorite foods, fried chicken with potato salad, baked beans, etc., cold drinks in our ice chest, fishing poles, swimsuits. This time in particular we had gone to bed for the night and were awakened by loud thunder and the lake water rushing over our car. My dad jumped up, started the car, and was able to get us out of there safely, but our trip was cut short. Also in the winter, when we would get several inches of snow, Dad would hook our old wooden sleigh up to the back of tractor and pull us down snow-covered dirt road. Once my middle sister and I rode together, she wanted to steer this time, but when Dad swerved too much, she got a face full of snow; she wanted me in charge of steering the sleigh after that.

Chapter 4

The Innocent Road

We lived on a dirt road about one and a half miles from the highway, and we rode a school bus to and from school. One morning, looking out the bus window, I saw a red-looking puddle with a shoe in it. I did not say anything to anyone as in the back country, there are many strange things. After my day at school taking the bus home, when we got to turn off for our dirt road, there were many emergency vehicles and police who had us turn around as there were two ways to get us home. When my sisters and I got home, we were very excited to tell Mom what had happened. She sat us down and calmly explained that someone had been killed and they were found by the creek bed below where I saw the shoe. That's all we were told until I was older and found out the body was that of a man and he was from a big city, possibly followed here by some organized group that had killed him.

Another school bus memory was of my mother's parents, who just lived a one half mile from us, and their property connected to a nice creek we liked to play in when snakes were not too bad. We were able to ride our bikes there; there was also an old railroad track we would adventure to. One afternoon, riding the school bus home, we saw smoke coming from our grandparents; the big, two-story farmhouse was on fire. It was a total loss—house full of love and memories gone, but thank God, no one was hurt. We as kids were filled with fear that they would have to move and not live near us, but they bought a trailer house and moved it onto the house foundation.

CHAPTER 5

APPLE ORCHARD

My grandparents had an apple orchard outside the city limits of Lincoln, and it was my first job other than farming, I picked apples. We would have to carry a burlap bag with strap to fit over shoulder and side, climb a ladder for high branches, then pick apples and put them in the bag until it was too heavy, then empty it into a trailer attached to tractor that we would move from section to section. As I got older, Grandpa promoted me to the shed where Grandma, Mom, and others would watch a conveyor and pick out bad apples before putting them in wicker bushel baskets for market. I enjoyed the orchard and working with my grandparents and parents, also friends my age that were hired. There is a business in town called Apple town, and the apples, homemade jellies, home-canned products, even peaches when in season were sold there. In October of every year, a festival called the Apple Festival is organized, and craftspeople come from everywhere set up booths, tents, etc. and sell their homemade items with kettle corn, funnel cakes, hot dogs, delicious barbecue, square dancing, and good musicians. Such a wonderful time for all, especially children. The school lets the teachers and parents bring kids down, which is walking distance for a field trip. There is a parade with horses, old antique tractors, cars, etc. Lincoln, Arkansas, is a special place with a lot of variety for living either in town or the country.

A TANGLED QUANDARY

We also had a special event last year, 2018. The *True Detective* movie crew came here and filmed a segment of their season 3. We have a lot of 1980s vehicles here that were used in the casting. I have a 1986 Toyota pickup that I bought from my mother and she from her dad. It has been in our family for over thirty years. I contacted Sarah Tackett of the HBO about my truck, and she called with an opening, but my schedule and theirs did not come together.

CHAPTER 6

CHANGING SCHOOLS

I went to school in Lincoln for ten years until my parents met some people at a prayer meeting that were Seventh-Day Adventist. We had always attended Baptist Church in nearby town. After going to their church for a while, we decided to join, and my parents transferred my middle sister and me to their Adventist School in Gentry, Arkansas. My parents also leased their bakery so we would have an additional income. My mother, grandmother, sister, and I worked their early hours before and after school, and my dad continued raising chickens. My younger sister and brother stayed in Lincoln school with Dad there when they got off the bus. I was old enough to drive our paneled station wagon and made bread and pastry deliveries to surrounding town stores.

One morning my, mother using a machine with two steel rollers that would flatten the dough for pastries, got her hand too close to edge of dough, and it pulled her hand through—thank God it was not broken, but it was sore and bruised. I really enjoyed the bakery and working there, not to mention the homemade bread, donuts, and delicious cinnamon rolls the size of dinner plates with pineapple, cherry, and blueberry in the center. Dad decided it was taking Mom away from home too much, and they sold the bakery but allowed my sister and me to finish the year at the Adventist school. My dad let me take our blue 1964 Chevy truck, along with my sister, to school, and one day I was sitting in class when the principal came in and said I needed to come with him as my sister had been in an accident. He said she wrecked the truck. I said, "No way, I have the keys," but when I looked in my purse, they weren't there. She had taken them, and I did not know. I took off running to the store where he said the accident happened. A car pulled up beside me, and the principal said get in, "I will take you." When I got there, I saw the truck through a fence and on its side. I just knew she was seriously hurt. Someone called my name, and there she was sitting with the paramedics and just shaken up. I was so thankful to God everyone was okay because the truck ran on propane and had a tank on the back. When my parents got there, Mom was scared and Dad was mad. I did not get to drive for a long time.

It worked out for a while longer until my dad who, I have not mentioned yet, was being a heavy drinker had a serious disagreement with church people about his drinking and he made my mother leave the church and put us back in Lincoln school. I was never happy about leaving my friends at Lincoln High and especially not happy we would not graduate together. This put me a year behind my friends, but I was glad to be back, and I also got back together with my high school sweetheart.

Chapter 7

High School Sweetheart

William was a football player and two years older than me, but we had a lot of the same interests—farming, riding horses, fishing—and he lived close to us. He proposed to me in his senior year, and we made plans to have a family and raise the kids as we had been raised, on a farm. His plan, along with some of our friends, was to join the service and get loans for farms, but they joined National Guard and he, the Marines. I wasn't happy about it because I knew he would not be able to come home for prom or special occasions to share with me, as my friends' fiancés would, but he assured me that he would.

He graduated and went to boot camp, and I was chosen to go to Girls State. So excited and such an honor, but William was coming home for a visit at that time, so I told my mother and teachers I had decided not to go. I was also nominated to run for football queen, such a fun and rewarding occasion. It was especially exciting because my cousin and a good friend also ran. The parade took place on our town square. Riding on our 1970s cars and trucks, in beautiful long gowns with hair done, we felt special with everyone cheering and applauding for each of us. I did not get to be homecoming queen, but was happy for the opportunity and memories. William, after training, was sent to his base in Texas. We wrote a lot, but he was not able to come home for special occasions and had very little time to visit. I was okay with it knowing he would soon have his three years in and we would start our lives together.

Things were okay for a while, but his next visit, he told me he had decided to enlist for another three years. I told him we had talked about his career in the service and neither of us wanted to be career military. He had changed his mind, and we parted stressed. After thinking about it, I decided I could not be apart from him another three years and we could get married and I would go back with him. I told him what I decided, and we wrote to each other for a while, but he started writing back less and less. When I did get a letter, he was distant. I tried to get him to tell me what had changed, but he just said the service was stressful and he had a lot on his mind. Of course, I was excited to start our life together and felt he wasn't feeling the same. After a few months, I wrote to tell him I thought our interests had changed and asked if he felt the same? Regrettably, he did, so we decided to break up. My friends' fiancés' time in National Guard was over, and they had dates set for weddings. As I reached my junior year, I was needing space, and Mom and I were disagreeing. Dad with his alcohol had no patience with what he thought was teenager nonsense. One day Mother and I were having an argument, and Dad intervened, pushed me against the wall, hands around my neck, and Mom had to get him off me. Seeing him raise his hand to her made me realize I had to get out of there before someone was hurt. People say in toxic relationships like theirs, my mother should have left Dad sooner for the sake of my siblings', mine, and her own safety. My opinion is, if she had left, we would not have a younger sister or brother, or their families, and I can't imagine our lives without them.

A TANGLED QUANDARY

CHAPTER 8

THE BREAKUP

A boy in my class had been asking me out, and I said yes. We dated through the summer, and he proposed to me, and we married in my senior year. During my senior year, I was nominated to write an essay about "What Is Right about America" and to possibly run for Miss Teenage Arkansas, but with my new role as wife, I did not continue this opportunity. Our lives were okay. We did not have as much in common. He was a guitar player in a local band, liked drinking and partying a lot, and I did not. I was not happy and knew my decision to marry him had a lot to do with my dad's rage and alcohol abuse. Also I was still in love with William. People with addictions—whether alcohol, drugs, pornography, in serious need of anger management, etc.—think they are only hurting themselves, but there are decisions loved ones are influenced to make because of the addicts' raging temper or abuse, effects all involved. My father and his alcohol addiction and raging, out-of-control temper caused Mother, my siblings, myself, and others to make decisions that ended up not good for us.

After a couple of months, I became pregnant but had miscarriage a few weeks later. I had been bucked off a few horses and participated in rodeo events, which included riding steer for participation in FFA rodeo queen competition, which I did not win. But I woke up from a hard fall to my mother and emergency techs standing over me. I was not pregnant during these events, but they could have played a role, along with my dad's rough abuse, in my not being able to have children.

After my miscarriage, my husband and I fought a lot and separated, later divorced. We did get back together a few times, trying to make it work, and I thought I was pregnant again, but it was false labor. I finished my senior year and graduated.

A friend of Williams told him I had divorced, and he came home for a visit to see his mom, brother, and stepdad. He called me, and we went out, rekindled our relationship, and married in a small church in Oklahoma.

BRENDA RAMSEY EAGLE

CHAPTER 9

DISAPPOINTED AGAIN

I left my family for the first time and moved with him to Grand Prairie, Texas, where we decided to start trying to have children. Regrettably, I did not get pregnant and started working at a store on the base about four months later around Christmas. William was coming home later and later. One day he was supposed to pick me up from work but did not show up, and I had to ask a guy I worked with to take me home. He did not come home all weekend. I was so scared, and we did not have cell phones then. When he did come home, he would not say anything except he was too tired to talk and went to bed. The next day he told me that a woman he was dating before he came back to Arkansas and had broken up with found out she was pregnant.

I said, "Is she sure it is yours?"

He said, "Yes."

I was devastated but told him that I had married someone else at the time he was seeing her. I knew he would want to be a father to the child and told him we could still continue our marriage. He discussed it with her, but I could tell their talk was about them getting back together, and for a few weeks, he was distant.

After New Years a member of his family passed away, and we came home for the funeral. I told him that things weren't right between us and I was going to stay in Arkansas so he could deal with his other relationship. I did not tell William that I had been late six weeks and thought I was pregnant. I was glad I had not as I miscarried again. Much time passed, and I knew we were not going to get back together, so I filed for a divorce. I went to work at a chicken hatchery in Lincoln and stayed with my parents, feeling disheartened with myself that I had not been able to have a successful pregnancy. I could not help but think maybe that would have kept William and I together. In the seventies through eighties, the choices for trying to find out why pregnancies do not happen were few, painful, and very expensive, also scary without someone to share the experience with. I went to see a doctor while in Grand Prairie, Texas. He said he would need to do extensive tests, and with the problems William and I were having, I decided to wait until I got back to my doctor in Arkansas.

CHAPTER 10

NEW FRIENDSHIP

Work was good for me, and some people from Ohio had moved into a rent house of my grandparents on our road. They had several children. My Grandpa and the dad became good friends, and he told me one of the girls was my age and that he told them there were some job openings where I worked. Shiela applied and was hired. One day her mom was not able to take her home, so she asked me if I could, and we talked a lot, realizing we have much in common. It was the beginning of a wonderful friendship, and we decided to rent a place and move in together. Shiela and I became very close to each other's families and them to each of us; she had several older siblings still in Ohio—a younger sister and brother here in Arkansas, as I did, and we all were very close. One of my philosophies is, "Blood is thicker than water, but love is thicker than blood," and it came true. A few months later, we moved in together, and I had thought that after I moved out from my parents, they would get along better. My mother had plans of turning my room into an office as even with farmwork, there is a lot to keep up with and that with this, along with few other changes, Dad would be happier.

My hopes did not come true as several months later, Mom called to tell me she and Dad were separating. They later divorced, and my mother moved into a house closer to Lincoln City limits, and Shiela's parents, sister, and brother moved back to Ohio. The chicken plant was very limited with their wage increases, so we both started looking for a better job, and a factory in Fayetteville, Arkansas, hired both of us. We worked their several months until Shiela's dad called her from Columbus, Ohio, where her parents and their two younger children lived. He told her he and her mom were separating and asked if we could go there as he owned his own business and wanted to move it and her little sister and brother back to Arkansas. So we gave our notice and headed to Ohio in an older model white Buick.

Excited to be traveling together on our own without parents, we were traveling through Missouri faster than allowed and realized police lights behind us. Laughing until almost crying, I pulled over.

The officer came to the window and said, "Are you girls in a hurry? I have been trying to stop you for a few miles."

I said, "I haven't had this car long, and I think the speedometer is wrong."

Luckily, realizing we were not running from him, he gave us a warning. Back on the road, and almost out of Missouri, the Buick started to have problems, and I pulled into the first gas station. The owner was a mechanic and said it was the water pump and he wouldn't be able to get the parts until tomorrow. We told him we were on the way to Ohio and asked where we could spend the night? He was very nice, took us to the nearest motel, and said when he finished the car, he would come and pick us up. We were very relieved to be safe and walked to nearby restaurant for dinner. As we were walking back, just before dark, we noticed two men following us. We got to our room, locked door, and started getting ready for bed when there was a knock at the door. Carefully looking out the window, I saw it was the two men. We turned the lights out and ignored the knock; soon they went away.

Later, after we went to bed, the men came back and tried to get in the door. We tried calling the office but got no answer. We quickly got dressed, crawled out the bathroom window, and ran to the restaurant that was still open and stayed there until the police came. They took us back to our room and kept an eye on our room until morning. The mechanic came and got us, and we were thankfully back on the road again. We made it to Columbus before nightfall and were safe with her dad.

A TANGLED QUANDARY

We were there for about a month closing business, packing, and getting the kids out of school to head back to Arkansas; her mother stayed in Ohio. We helped them get settled and the kids in school, then Shiela and I found jobs at a new factory that started up in Prairie Grove, Arkansas. The factory made small-horsepower motors for garage doors, chicken house fans, etc. Many good people and relationships were started, including a young man I cared about and I started dating. He had a close friend that he introduced to Shiela. They dated a few months, and he proposed to her. She then became pregnant and moved in with him. Things between them were fine for a while until she started suspecting he was into some things she wasn't comfortable with or wanted for her child. She asked him several times about it, and he denied and said she was imagining it, or the things she was finding belonged to someone else. Finally, after a month or so, she called me to meet her for a talk and was serious that she had to leave him. My family and I supported her; we moved back in together as my boyfriend and I had broken up several weeks before and I had moved in with my mother, sister, and little brother.

Her baby daddy just dropped out of her life, and she asked me to be her Lamaze couch. I was honored, and we attended classes to prepare for the birth; she found out she was having a baby girl. The Lamaze classes were good for us to feel more in control of the events that would happen, but there was so much information I took seriously that I would need to know but didn't—like a tennis ball for Shiela to squeeze that she did not use (she used my hand instead, which still had the bent ring I forgot to take off); also the Jell-O she nearly threw at me for asking if she wanted it; music she wouldn't listen to—but they were good distractions for us as her labor was long, and the breathing exercises did help. After her daughter was born, she had a ceremony planned with her sister-in-law and her mom, who was a pastor of a church in Lincoln.

During the dedication of Lexi, she asked me to stand with her and was asked to be her daughter's godmother. Again I was honored and said yes. It is such an amazing bond between mothers and their babies, and they will always be their babies no matter their age. Just like the fingerprints being formed when they are in the mother's womb, by pressure on the fingers against the mother's womb lining along with their surroundings, create what is called "friction ridges," the faint lines you see on fingers and toes which can only identify *you*. Lexi was born in my heart, not under it; she is and will be like my child no matter our situation "always and forever."

CHAPTER 11

LIVING BACK IN ARKANSAS

My entire family and Shiela's have become very close, and we love them both as our own, and that unconditional love will last forever.

We have had so many good times and have seen each other through the bad times. My mother, Nanny to so many children, along with farming, worked out well together, and Shiela's daughter, Lexi, stayed with my mother until she started to school. Also Shiela's mom and dad divorced, and my mother and Shiela's dad moved in together about a year later. Shiela's dad was a very special man, and I loved him very much and was so hopeful he and my mother would have a very long and happy life together. But things and people who did not feel the same kept it from happening: Shiela's dad has passed away.

Lexi is now married and has three children of her own, which has given me the chance to experience being a grandmother. Mother has also taken care of her children, who are now all in school. I am so thankful for my friend Shiela and her daughter for allowing me to experience some of the joys of parenting and grandparenting. I will always cherish those years. Along with being involved with my brother, sisters, and their children, they are all such blessings and bundles of inspiration. My mother has always been a mountain of support to me and a good friend. I have also been very blessed to work with very good family-oriented people and have become close friends / family especially my dynamic 6, who have faithfully stood by me through some very trying times. We love, laugh, and live a lot. There are also a group of girlfriends that I am proud to call mine who have stood by me since my childhood years and in the factory I worked at in Prairie Grove; there is strength in numbers. Two of these friends have lost adult children and even though their loss will always be with them they have continued to be wonderful and encouraging parents and grandparents, I have such respect and admiration for them. Another of these beautiful women had two serious health issues that could have ended her life, but she instead came back better and stronger that ever before. There is one girlfriend whom I have shared my burdens with many times, and she has always come through for me and helped me see things in better perspective as well as encouraged me when I barely had enough enthusiasm to move. I will always be so thankful for you, BKA, and I hope you know I am and will always be there for you. I also have two male friends who have shared my dramas for over eighteen to twenty years. Also my brother, who is an awesome man, husband, and father, for getting a man's perspective helps in a lot of situations. I am so blessed that I have, after many years of grief for not being able to have children, realized my purpose has been to be a mother or supporter to adults and their children. As for most of my sisters' lives, they have been single parents; it takes a village, it is so true. During this time, my dad remarried, and my stepmother had children, adding to our family tree.

 Shiela met a man in her church, and he fell in love with her and Lexi. They married that summer and were very happy for many years. I was working for the US Postal Service, and a friend introduced me to a man she had known that was divorced with two sons. This relationship was so disappointing as he had only been married once, with two grown sons who had families, good jobs, and respectful of all that they had. On the outside looking in, anyone would think that besides everyday, normal problems, he was a good, family-oriented man.

 We dated for a few months. He proposed, and we married in the fall. We were married for three years, but not long after we were married, I started noticing his personality changes and paranoia. He was drinking, but it was vodka. I couldn't smell it, but I was finding the empty bottles. This was an even bigger problem because he was mixing it with a prescription drug he had to take for anxiety, and he became paranoid. He would watch for me to come home from work, and if I came from a different direction or had dirt road dust on my car, he would accuse me of seeing someone else. I was a mail carrier and had deliveries on dirt roads and also had family I might go see that lived on the other side of town. Also the grocery store was on the other side of town. He was a corrections officer for the state and had suffered a work-related incident that required him to take the drug. It was Ativan, but of course, he did not tell me about it until after we were married. No matter how much I cared about him, I could not help but feel trapped.

After talking to him about it, I found out he had been on it for years, which, after researching, sounded ridiculous to me. I felt he should not be on it anymore, but he was very serious he should be, and when I insisted he come off it, the withdrawal symptoms were horrible. We separated, and I told him we had to get some help for him, but when the specialist told him what it would involve, he refused and said he could deal with it on his own. He would have been able to get an honorable discharge, but he did not want to rock the boat. Several things happened—such as, a fellow mail carrier came to me and said, "Don't be alarmed, a neighbor called the police, your husband is locked in his truck, and they cannot get him to wake up." Another time he came home drunk, sitting in the dark in living room, when Lexi, whom I had for the weekend, and I came home. I could tell he was drunk, so I took her home. When I got back, he was gone, so I locked the door. When he came back, I told him to sleep in the truck, but he broke the door open and shoved me against the wall. My mother, who was living in an apartment next-door, called the police. When they got there, they asked me if I knew where he kept his gun. I said yes, and they handed me his keys and said, "Don't let him drive," and left. We separated, and I moved in with my sister.

One morning I woke up to find him passed out in the back of his truck, his clothes soaking wet as if rained on because of the horrible withdrawal symptoms to try to come off the drug on his own.

That was the last straw, until he took the situation seriously. But he insisted it wasn't a problem, so we divorced a few months later.

I just could not go through what my mother had with my father. Life is too short when it comes to relationships to waste it in the margins with inadequate people who don't have respect for their partner's life, or even themselves, enough to work together.

Thankfully, I still have my family and friends that I am very involved with.

CHAPTER 12

MY NEW CAREER

I enrolled in Northwest Arkansas Community College to get my patient care assistant and CNA certification. After completing college, I applied to a home health agency to work with physical therapist to teach range of motion, bathing, dressing themselves, walking, and staying with patients while family worked or just needed to run errands. I had many patients after eight years of working with veterans, all types of men and women with many different conditions. Being able to form relationships with them and their families meant so much, but as with many other things that are good get changed, it got changed by the new health-system rules. My line of health care was cut, and the only outlets were hospitals and health care / nursing home facilities, which required twelve-hour days and using lifting equipment or our own physical lifting skills, and at my age, they were too hard. Certified nursing assistants and nursing aides don't get the credit they deserve as besides the long hours on their feet, they see and hear a lot that is reported to their floor nurse and save many lives. Also the heart-wrenching loss they feel when even though they are told not to get close to patients, they do mourn. Even after changing my career, I hear from the loved ones of patients that invite me to weddings, graduations, and let me know their family member had passed away.

I heard a baking company in Gentry, Arkansas, was hiring, and I applied, was hired, and started two weeks later. Popular snack cakes is their business. There were many different jobs to do with this process, besides eating them when on break, as they kept the break room highly stocked with a very good variety of snacks. People of all ages and lifestyles worked there, and for the most part, we got along well. Also the company had strict rules about cleanliness and freshness, which showed in the purchase of the products. I am very blessed to have formed relationships while there that I still enjoy even after changing jobs. We have our annual Christmas party with a gift exchange from names we draw at the end of every party to have a year for considering different places we might shop. We also plan a summer vacation together to catch-up and celebrate our birthdays, I call them my dynamic 6. We are always there with love and support, but also if we see one stepping out of their comfort zone as into a toxic relationship or something that's not themselves, we intervene in group form.

CHAPTER 13

WILLIAM'S LOSS

One day a close friend from school called me to say that there was a Lifetime Movie called *Loves Deadly Triangle: The Texas Cadet Murder*—and that it was about William's daughter's murder. I could not believe it, and when it finally came on, I taped it, which was beyond my comprehension. A brutal murder of a sweet young woman in her high school years, after having a relationship with a young man that she did not know was engaged to someone else, and the couple conspired to murder her after he confessed the affair. After I watched the heart-wrenching movie, a show called *Beyond the Headlines* came on about their arrest and trial. A segment at the end showed her mother, classmates, etc. planting a tree, then her tombstone at their cemetery showing her birth and death. I then realized it was the child of the pregnancy that he found out about when we were married. So many feelings ran through me—that could have been our daughter. After several months, the reality set in that as sad and disappointed as I have been to not have had children, I would rather not have had than to have lost one as he had. I just wanted to talk to him and tell him how sorry I was, so I decided to write this book in hopes that someday he would read it and know my sadness for him. One Sunday, July 8, 2012, at 11:30 a.m. our time, the phone rang; we had a landline at that time.

 I answered and said, "Hello."

 He said, "Hello, Brenda, do you know who this is?"

I hesitated for a minute, but his voice was the same. I started to cry but said, "Yes, William, is that you?"

He said, "Yes, how are you?"

It had been over thirty years since we had spoken to each other, but feelings of it like just being yesterday came rushing back. I was in shock; so many things were running through my head.

"I am fine, how are you?"

It was 9:30 p.m. in Afghanistan. We talked for at least an hour, and he told me he had lost a close friend who had died from enemy fire there and realized he might not have much time left and thought of me. He wanted me to know how sorry he was for our broken marriage but that he really had and still loved me. I told him I was sorry also and that the only way I knew to let him know was to write this book. He was excited and gave me permission to do so. I asked him about his daughter. He was sad that I had found out but said it was horrible, and I could tell by our conversation that he had not come to terms with it as his daughter's mother had. The movie showed her planting a tree for their daughter and saying in their trial not to give them the death penalty, as they didn't want their parents to suffer the loss she and William had, but neither time did it show him with her; they are divorced. He has two sons, a nephew he has helped raise, grandsons, and a granddaughter; he is also remarried. He has retired as a Marine but is still in Afghanistan as a civilian working for a company that takes apart FOBES to ship back to the States, and they do come under fire quite often. He called me several more times in the course of a few months, and always on Sunday. I truly enjoyed our conversations and catching up. Once my mother was here, and she talked to him a little while. The more he called, the more I began to wish we could see each other, and then the reality that he was married and already had a life with someone else came to mind. I thought about it in between our conversations and decided to ask him if he was happy with his wife. He said yes, so I told him it had meant more to me than words could say that he had contacted me but that we should not continue as it was starting to bother me that I was feeling too close again. He is the only man that I have cared for that still gives me flutters in my stomach when think about him, just like when in high school.

    I told him to let me know his address so I could send him a copy of this book, and he did call me once more and left a message that I could call him at certain times of the week as he was home for a while, but when I called the number, it said the message system was full. To have him acknowledge that he had loved me was such a reassuring thing as, for so many years, I felt was it possible our love was one-sided. To know that at a time, when he was in a serious situation and might lose his life, I was on his mind will always mean more to me than anything else he could have said.

CHAPTER 14

TRAGEDY AGAIN

There is also another tragedy of a man I dated and cared about. He was the friend of Lexis's father, who lived in Prairie Grove, Arkansas, named Ronnie. We dated for almost a year but disagreed a lot, and not having same interests, we broke up but remained friends. He was so sweet-natured.

I saw him in the local Dollar General store a couple of years ago, with a cute young boy. I asked him, "Who is this?" I knew the little boy wasn't his blood child as he told me when we were together he could not have children.

He said, "My reason for living, and he has made a big change in my life."

The gleam in his eyes and the big smile on his face and the boy's told me he meant it. A year later, on the local news, I heard his name. He had been found on a cold winter day, his lifeless, shirtless body on the side of a road. The news said a carload of people had driven him around all night, beating him, while he begged them to take him home. But of course, they did not. I had been told by friends and also other news reports, the local paper, that a few people suspected something was going on in the place where they stopped for snacks and gas. I so wish they would have called police, even if it turned out to be nothing. We as a new world are full of so much violence and have got to be more vigilant.

A TANGLED QUANDARY

I often think of that day I saw him and the young boy and wish I had taken more time to visit with them.

"Life is not measured by the breathes we take but by the moments that take our breath away."

CHAPTER 15

WHERE I AM NOW

Shiela's husband passed away, and she has remarried and moved to another city. Lexi and her family have moved to put her kids in a bigger school, and I don't hear from them very much anymore. There is something I will never understand when we fall in love with someone as our own. How can someone just come into their lives and tell them or make them feel as though they should not love them because they're not blood family?

For example, children not born to but grow up in a loving family, get in a relationship or married to have the new family make them feel like they are silly or the love is not normal because they're not blood. This to me is so disappointing, and I will never be bullied into not standing with my loved ones in every life involvement.

My mother, even with the heartbreak of my dad getting remarried, took care of his wife's kids and grandkids as their nanny; that's just the kind of family we are. Have you ever noticed a herd of mama heifer cows with their baby calves? They all watch each other's babies. I have even seen a heifer letting another baby nurse, and while the herd is grazing in another part of the field, they put all the calves in one place with one or two mamas keeping watch. What a wonderful thing the animals try to teach us, like the cover photo of two horses grooming each other.

A TANGLED QUANDARY

CHAPTER 16

CONCLUSION

I have changed jobs to a high school cafeteria employee and school bus driver, which at this time of my life is very rewarding. Deciding to write this book has been the best idea for me that I have had in a long time. I have kept a lot that I am writing about inside all these years, feeling less of a woman not being able to bear children. With the added anxiety for being a women in the 70–90s girls felt pressure to be a wife and mother above any other hopes or dreams.

I went through so many emotions, like envy and jealousy toward my sisters and brother for being able to have children, and then quilt for being so resentful.

But then throughout the course of my life, and much faith in God and prayer, I have realized that the rewards were much greater in developing a loving relationship with them and my wonderful nieces and nephews.

I am also hoping to reach people with similar circumstances with not being able to have children, having parent and family issues, dealing with addiction in relationships, and disappointments, as there are a lot we can be involved in and make our place in this world with many other rewarding situations. Is this not what our life was meant to be? Maybe not to stop and hide our happiness but to extend ourselves, time, and God-given talents to other realms of life for those who might be too busy to get involved in. Just not giving up or closing our hearts to people who would accept our different ways of finding happiness and being involved with others who have the same circumstances. Also, throughout my life, an organization that has really helped me is Al-Anon. The support of the good and caring people there showed me that loving people with addictions is not wrong, but we should not allow it to consume, afflict, and definitely not abuse us, our lifestyles, or other loved ones, and to accept their addiction as an illness/disease.

I also wish my mother would have had this organization and support when she was dealing with my dad. This is not all I intend to write about. There are other issues, like people with addictions thinking it only affects them, parenting and communicating with children as they are so near and dear to my heart. I feel that at times they are so misunderstood, especially in this day and time with so many changes and safety issues, children and young people need to be able to talk to their parents freely without fear of consequences.

I know that the statement "We should parent and not be their friend" is supported. But do we, as adults, choose to talk to people we feel safe and comfortable with or someone that would talk critical of our thoughts and fears, even make fun of us? Like the saying "Don't hide your light under a basket," also use their feelings of being misunderstood and not being able to communicate. These feelings with their parents, grandparents, aunts, uncles, etc. can make them more vulnerable to be lured into horrible situations with the wrong people.

I watched a documentary on sex trafficking, and these things are exactly the ways children and young adults are preyed upon; especially with social media, they tell all to whomever will listen.

More to come...

ABOUT THE AUTHOR

Brenda Eagle was born in Denver, Colorado, the oldest of four children—three girls and one boy. Her parents moved back to Arkansas when she was eighteen months old. She was raised on a farm and attended Lincoln public school where she graduated, and then she worked to put herself through business college. My dreams were to marry a young man who shared the same interest in farming and wanted children to share it with. After marrying her high school sweetheart, she found out she could not have children, and her life and ambitions changed. It took her a long time to realize there were often areas in her life that she could be happy and useful in, especially after her high school sweetheart suffered a horrible tragedy when it looked like he had gotten what he dreamed of. Her goal of this book for you who read it is that she hopes you will not be afraid to make new dreams and realize your life is very important and worth the effort. Life is measured not by the number of breaths we take but by the moments that take our breath away.

CPSIA information can be obtained
at www.ICGtesting.com
Printed in the USA
LVHW041256101220
673819LV00020B/508